Anne Abrams:
Engineering Drafter

Published by
Twenty-First Century Books
38 South Market Street
Frederick, Maryland 21701

Text Copyright © 1991
Jennifer Bryant

Photographs Copyright © 1991
Pamela Brown

All rights reserved. No part of this book may be reproduced or utilized in any form or by any means, electronic or mechanical, including photocopying, recording, or by any information storage and retrieval system, without written permission from Twenty-First Century Books.

Printed in the United States of America

10 9 8 7 6 5 4 3 2 1

Cover and book design by Terri Martin

Dedicated to all of the working moms who helped with this project

Library of Congress Cataloging in Publication Data

Bryant, Jennifer
Anne Abrams: Engineering Drafter

Summary: Portrays a day in the life of an engineering drafter who is also a busy mother.
1. Drafters—United States—Biography—Juvenile literature.
2. Working mothers—United States—Juvenile literature.
3. Abrams, Anne—Juvenile literature.
[1. Abrams, Anne. 2. Drafters. 3. Working mothers.]
I. Brown, Pamela, 1950- ill. II. Title. III. Series: Working Moms.
T353.5.B79   1991   604.2'092—dc20   [B]   [92]   90-24378 CIP AC
ISBN 0-941477-51-7

Working Moms: A Portrait of Their Lives

Anne Abrams: Engineering Drafter

Jennifer Bryant
Photographs by Pamela Brown
Photographic Consultant: Bill Adkins

TWENTY-FIRST CENTURY BOOKS
FREDERICK, MARYLAND

It's 7 o'clock in the morning. The bright sun is shining on the Brandywine River . . . and waking up Anne Abrams. She has another busy day ahead, and it's time to get ready.

Anne Abrams is a working mom.

Anne is an engineering drafter. She makes drawings that are used to build houses, roads, bridges, schools, and other structures. It's an important job. The men and women who work on construction projects must carefully follow the directions on drawings like those Anne makes. Her technical drawings are a kind of map that shows construction workers what to do. They depend on people like Anne for a detailed and accurate guide to the work they have to do.

Many other people depend on Anne. The families who live in the neighborhoods she helps to plan, the kids who play in the parks she helps to design, the people who ride on the roads she helps to build: they depend on Anne.

Anne is also a mom, and her family depends on her. That's an important job, too.

Anne takes a moment to think about the day ahead. "It's going to be a tough one," she says to herself. "I've got to finish the plans for the new housing development."

She glances at her watch. It's 7:30 A.M.: time to wake up the boys and get them ready. Mornings always seem to be such a rush! Anne has to dress Sam and get breakfast ready. Sam is two years old. His brother Alex is three. Alex can dress himself now, but he's awfully slow about it.

"Let's hurry up," Anne calls. "There's a lot to do this morning."

Anne wishes that her husband Michael could be here to help. But Michael has to leave the house very early in the morning to get to his job on time. When Michael leaves, the house is dark, and Anne and the boys are still asleep.

The last button is done on Sam! The boys race to the kitchen with Anne close behind. She helps Sam into his highchair and ties on his bib. Just before Sam settles into his seat, Alex gives his little brother a playful pinch. "Boys," Anne says, "please sit down and eat your breakfast!"

Another day is under way for the Abrams family.

Anne turns to look out the kitchen window. "For just a moment," she says quietly. The Brandywine River bubbles over the rocks in the bright, morning sunlight. It flows under the bridge that crosses nearby Devereux Road and past the woods that border the Abrams' house.

Anne is glad that the water and the woods are her only neighbors. She watches two mallard ducks land smoothly on the river and swim toward the mossy bank. "It's such a peaceful scene," Anne thinks.

"Mom, I'm done!" Alex's voice suddenly interrupts his mother's thoughts.

"Eat a few more bites," Anne says, "and then put your bowl in the sink."

Anne takes a last look at the Brandywine. "No time for daydreaming this morning," she thinks. Anne has been given a big assignment by her boss. She must present the final plans for the new housing project to a meeting of the local town council. She has to know every detail of the project. She has to make sure her drawings are perfect.

But right now Anne has to make sure that the boys are ready to go. "It's the morning rush hour," she says to herself. As Anne turns away from the window and the Brandywine, she remembers when Sam and Alex were babies and she stayed home with them every day.

Mornings were not so hectic then.

It's 8 o'clock. The boys have finished their breakfast, and the Abrams are just about ready for the day. But first there are a few more mouths to feed.

Anne makes a quick visit to the back-yard chicken coop, where she feeds the family's four hens. "The girls," she calls them. "And what a noisy group of girls they can be!"

"Most people are surprised when they find out we have chickens," Anne says, "since we don't live on a farm. But the girls are very happy here. They're a regular part of the Abrams family."

Finally, the boys are ready, the pets are taken care of, and it's time to go. But where's Sam?

"He's looking for bugs," Alex tells his mother. "To take for show and tell."

"I'm going to be late for work," Anne says to herself. "C'mon, Mr. Bugman," she calls out to Sam, who is busily searching the front yard for a worm or maybe an interesting beetle. "We're off!"

First stop: the Little People Day Care Center, where the boys will stay for the day. Anne can remember how hard it was to leave the boys at the day-care center for the first time. "I felt awful about not being with them," she recalls. "But after a few weeks, I realized that Alex and Sam were really having a good time. They like the activities here and enjoy seeing their friends."

It's still hard to leave the boys. "But it's a lot easier for me to go to work knowing that they are happy and in good hands," Anne says.

Anne drops the boys off at 8:30 A.M. She gets a good-bye kiss and hug from both of them.

"Keep an eye on your little brother," she tells Alex.

Second stop of the day: the Glenmoore General Store for a cup of coffee. Anne parks her car under the sign that reads, "Sandwiches, Groceries, and Country Gifts."

Glenmoore is a small, friendly town just a few miles from the Abrams' home. The main street is called Creek Road because the Brandywine River flows close by. There's a fire station. There's a post office. There's a gas station.

And there's Mr. Carr's General Store.

"Good morning," Mr. Carr says when Anne walks in. Mr. Carr is the store's owner. He has lived in Glenmoore for years and years, and he knows everything that's going on. He keeps Anne up to date on the latest town news while she pours herself a cup of coffee. But there's not much time to chat today. Anne has to be on her way.

"You have a nice day now!" Mr. Carr says.

The car radio keeps Anne up to date on the rest of the news. "I hardly ever have time to read a newspaper," she says. "The radio tells me what's going on in the world."

Anne works in a busy town not too far from her home. The drive takes about 20 minutes. "It's one of the quieter times of my day," Anne says with a smile.

Anne gets to the office at 9:30 A.M. She goes right to work. There's a lot to do today.

Have you ever wondered how a building gets built? Or a road? Or a neighborhood park? How do the people who build our homes and streets, our schools and hospitals, know how to do it?

It may seem like magic, but it isn't. The builders follow a construction plan.

Anne is one part of an engineering team that designs construction plans. But how is a construction plan made?

The first step is to learn as much as possible about the land where the building or road is going to be placed. This is the job of the surveyor. Surveyors measure the land. They make a map of the land showing the way it is shaped, how it slopes and curves. They also mark off property boundaries.

Surveyors use special instruments like a transit, which looks like a small telescope. They also use different types of measuring rods and chains. They must take very careful notes. The survey measurements are going to be used by the engineering team to make a construction plan.

At this point, a project design has to be made. This is the job of the engineer. Engineers supervise the building of construction projects. They study the survey measurements and decide what shape a project should take.

Anne's job is to make a drawing of the engineer's design. But not just any kind of drawing. It has to be a very accurate and detailed sketch of what the construction project will look like. It is the map that the men and women who build the new project must follow.

For the past few weeks, Anne has been working on the plans for a new housing development. This is a special project for Anne. The development is going to be built right down the road from the Abrams' home. "When it's finished," she says proudly, "I'll be able to drive past the new homes that I helped to plan."

Anne turns to the drawings she has been working on. The builder owns 30 acres of land and wants to build 20 new houses.

But he can't just put the houses anywhere he likes. There are many questions to answer. Where are the best spots for the houses? How many houses can be built on this land? Will there be any playgrounds or recreation areas? Where will the roads go? What about the sewer pipes? What about electricity and phone lines?

The builder needs more information. He needs a plan before he can begin the construction work. That's why he hires a company like the one where Anne works.

Anne does her work at a large, tilted drawing table. It's called a drafting table. She uses a variety of instruments to make her drawings: a compass, a protractor, and special rulers to make angles and curves. Her drawings are made on stiff, plastic film called mylar.

When Anne begins a drawing, she uses a "no-smudge" pencil, so if there are any mistakes, they can be easily erased. Later, when the drawing is complete, she goes back over each line with a black inking pen.

Like a mapmaker, Anne draws "to scale." "To scale" means that a small measure (like an inch or centimeter) on her drawing represents a larger measure (like a yard or meter) on the project. So Anne must be very sure that the drawing is precise. Even a small mistake on Anne's part could mean a big mistake at the building site.

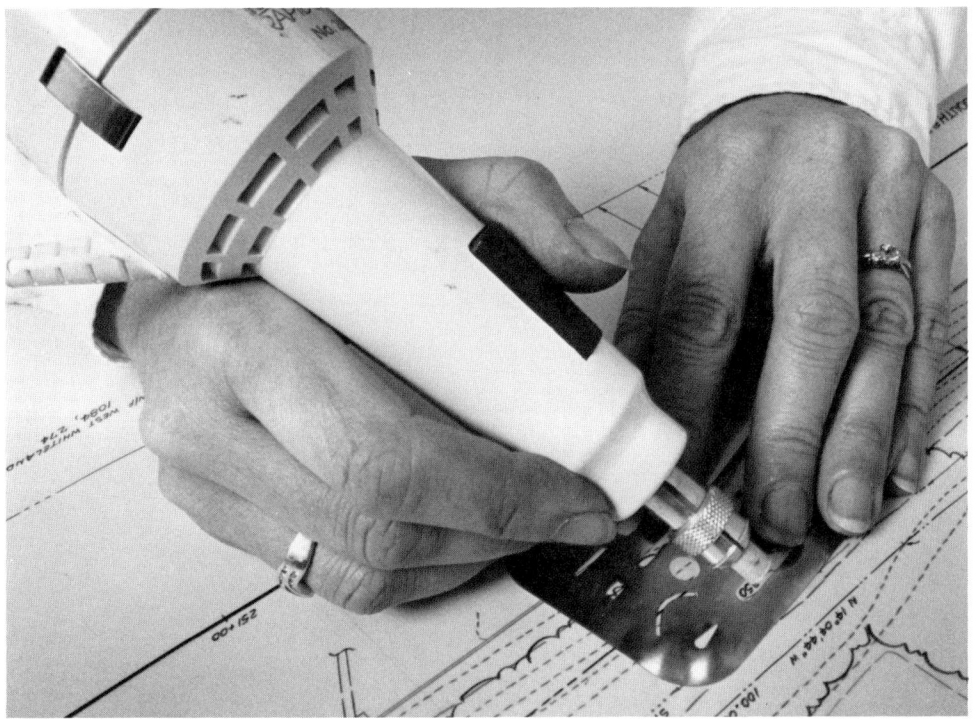

"That's why neatness and accuracy are such important qualities of a good drafter," Anne explains. "A drafter has to communicate technical information clearly and exactly. There's no room for error."

Anne prepares a drawing called a site plan. She may prepare many different site plans for the same project. These plans show different views and parts of the project.

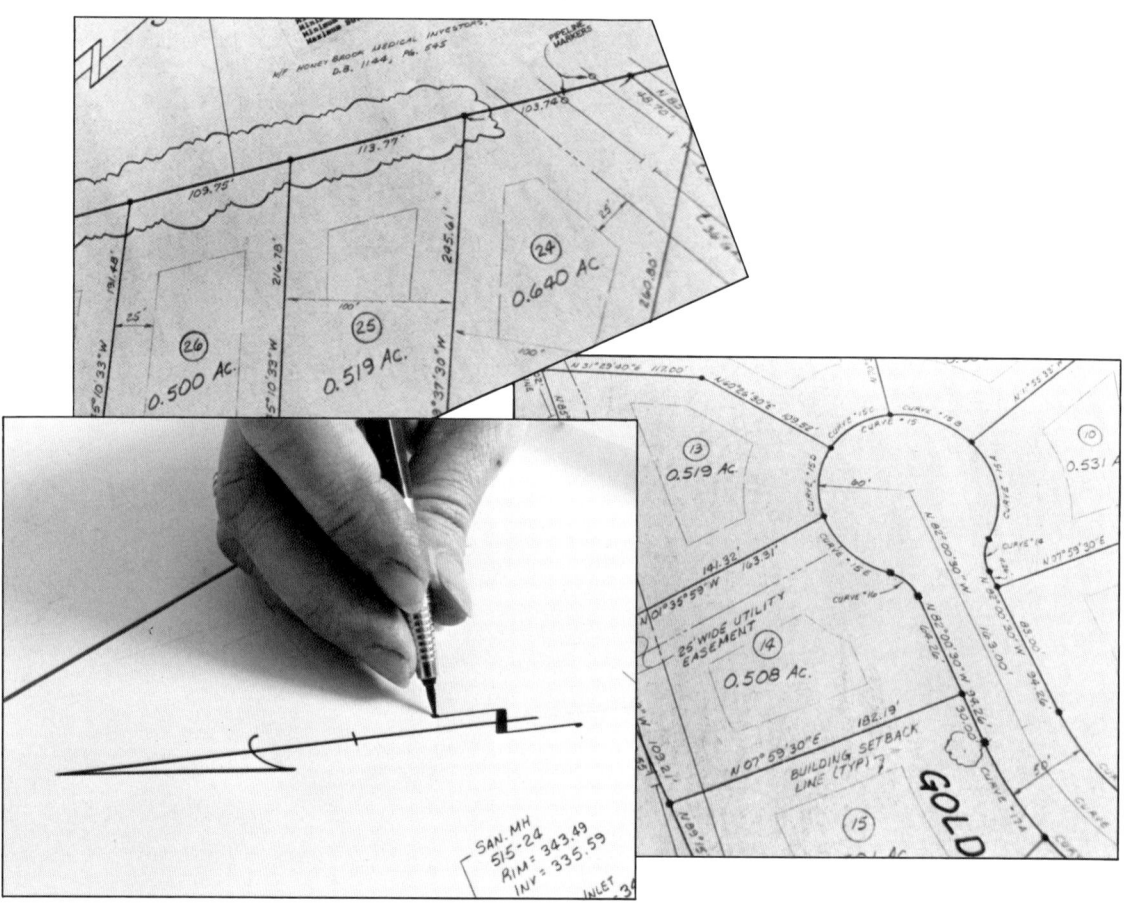

Anne puts a "North" symbol on her site plans. This is how she marks her work. The "North" symbol is Anne's personal signature, and she's proud to place it on her work.

The site plan is copied onto special paper. The lines on the final drawing are blue. That's why the final drawing is called a blueprint.

Anne applies a stamp bearing the name of her company onto each blueprint.

Designing and drawing a construction plan require a team effort. Anne must work closely with the rest of the engineering team.

Sometimes Anne needs to visit the building site. She may want to review the survey measurements. She may want to talk to the engineer about his design ideas.

Anne has already been to the site of the new housing development several times. She goes to check every detail of her drawings.

"This project has been a real challenge for me," she says. "The land here is very close to a flood plain." A flood plain is an area, usually near a river, that may be flooded in very rainy weather. "I had to be sure that there would be enough room between the houses and the river."

Step by step, Anne's drawings become more complete. She draws the houses and yards. She draws the roads and sidewalks. She draws the landscaping (the lawns, trees, and shrubs) and playgrounds.

Step by step, the housing development comes to life.

The kind of technical drawing that Anne does is demanding work. It requires concentration and attention to detail. It's 10:30 A.M.: time for Anne to take a break. Leaning back, she looks at the picture Alex made for her just the other day. She keeps it now on her office wall.

Anne smiles as she remembers asking Alex what he wants to be when he grows up. "A baseball player!" he replied confidently.

Anne remembers when she was a child. She dreamed of becoming a veterinarian. She loved to read about animals. She loved to take care of them.

Drawing was another one of Anne's favorite activities. She used to sketch outdoor scenes, like the lovely woods behind her home, or the gentle deer and playful squirrels that lived there.

She also liked to watch things being built.

"As a child," she recalls, "whenever I saw workers building a house or a bridge or a road, I would stop to see how it was done. My father was an engineer, and sometimes he would take me to his construction projects. I would listen carefully as he talked to the builders about each detail of the project. Since then, I've always been fascinated by how things are built."

But choosing a career for herself wasn't easy.

"Somehow, by the time I entered high school," Anne says, "I was really confused about what I wanted to be when I grew up. I was interested in a lot of things. But I didn't have any specific career goals."

After she graduated from high school, Anne worked at several different jobs. But she just couldn't find anything she really liked. "Something was always missing," she recalls. "I wasn't learning any new skills. I wasn't challenging myself."

Then, one day, she stopped to watch a group of surveyors at work. They were marking off the boundaries of a new shopping center. Anne looked at the land being marked off. Where the vacant lot stood, she imagined rows and rows of bustling stores. She tried to picture how the shopping center would be built.

"What a great job," Anne thought. And the more she thought about it, the better a surveying job sounded.

That very day she decided to contact a local technical school for more information. "I asked about the school's training programs."

She remembers the surprise of the school's director. "Well," he muttered, "women don't usually do this kind of thing."

"That just makes me more determined," replied Anne. She signed up right away.

Anne was the first woman in the drafting program. She had to be *very* determined to choose such a career. "It was scary sometimes, being the only woman. I felt out of place."

"But I knew I could do it." And Anne did. "But I'm glad," she adds, "that things are different for women now."

Today, young women are encouraged to develop their technical skills. High-school courses like mechanical drawing used to be "Boys Only." But not anymore. More and more girls are trying out technical jobs—and finding out that they like them.

Both men and women have to go through the same kind of training to become a drafter. After high school, it is necessary to complete a drafting program at a technical school. Like the rest of the people in the drafting program, Anne took courses in mathematics, science, surveying, and technical drawing. On-the-job training is also very important.

It wasn't long before Anne got just that kind of training. A few weeks after Anne graduated with a drafting degree, she was hired by a surveying company in Boston. Not only did Anne practice the surveying techniques she had learned, she also gained practical experience in technical drawing.

"My surveying courses were a big advantage for me when I became a drafter," she says. "I understood what this kind of drawing was all about. It was more than lines, numbers, and angles. They represented real things in real places."

Anne and Michael were married soon after she started her first job. Two years later, a change in Michael's job brought them to a new home in Pennsylvania. But Anne was fortunate. Her drafting skills were in demand there, too. She was hired by the same company she works for today. At that time, though, the company was just beginning. There were only two surveyors, and Anne was the only drafter.

"I did most of my work on my drafting table at home," Anne recalls. "I only needed to work part-time to finish my assignments, but that was okay with me. I used the extra time to help Michael fix up the rooms in our new house."

Anne and Michael took special care with the spare bedroom. Soon there would be a new baby in the house!

"It was a big adjustment for us when Alex was born," says Anne. "We were very happy, of course, but there wasn't much 'free' time anymore. Somehow, though, I managed to do my drafting around the baby's schedule."

And there was even less free time a year later, when Sam was born. Not only was the Abrams family growing fast, but so was the amount of drafting work Anne had to do. She could no longer work at home. "It was impossible to get anything done while I was with the boys," she recalls. So Michael and Anne looked for a good day-care center and found one nearby. "It was a happy place," Anne says.

"It's worked out pretty well for all of us."

Anne gets back to work. She spends the next several hours working on the site plans for the new housing development. She decides to have lunch (a cup of yogurt) at her desk today. She wants to check her drawings again.

At 1:15 P.M., she meets with the project engineer and the head of the survey crew. They discuss several aspects of Anne's site plan.

"I'm concerned about the number of storm drains," the engineer says. "Maybe we should add one more."

"If it rains heavily," Anne agrees, "one drain may not be enough."

Pointing to one of the playgrounds planned for the new neighborhood, Anne makes another suggestion. "I think more landscaping is needed here. Otherwise, we could have a problem with erosion."

Slowly and carefully, every detail of the site plan is reviewed. "It's not just a drawing," Anne says. "These are homes and streets and parks. Real homes and streets and parks. People are going to live here. Kids are going to play and grow up here."

"It has to be perfect!"

It's 3:00 P.M.: time for Anne to make herself a cup of tea. Sipping her tea, Anne tries to imagine a young family moving into one of the homes in the new development. But her thoughts wander to her own home. Did she remember to tell Michael to pick up bread and milk on his way home? When should she schedule Sam's medical check-up? Did anyone call the plumber about the leaky faucet? Who's going to take Alex to the dentist?

"There's always something to do or think about," Anne says with a sigh. "No, there's always *too much* to do and think about."

When Anne returns to her drafting table, she makes a few last changes to the drawings. Now she can sit back for a moment and look at the final site plan. "It was a lot of work," Anne says, "but I think it's a good job." Looking at her work, she once again tries to picture a family living in the neighborhood she helped to plan and design. This time, Anne can almost see kids running in the yard.

"I'm glad I'm a working mom," she says to herself.

It's 4 o'clock. Anne has finished her last drawing. She puts her drawing instruments away. It's time to go home.

Anne's work schedule has changed since she and Michael started their family. She works from 9:30 to 4:00 three days a week. "My hours are just right for my life now," says Anne.

"I'm lucky to have time off to spend with my family. Many working moms aren't so fortunate."

If one of the boys gets sick, Anne can stay home from work. But sometimes it's not so simple. There was the time Alex had the flu. He was in bed for four days. "I had a 'rush' project to finish," Anne recalls, "so Michael stayed home with Alex for two days, too."

"I'm also lucky to have that kind of support."

At 4:30 P.M., Anne picks up the boys from the day-care center. But her workday isn't finished yet. Far from it.

"Being a mom is work, too," Anne says.

There's a lot that Anne has to do just to keep the house running smoothly. "That's what days off are for," she says. "I take the boys with me to do errands: to the grocery store, to the bank, shopping."

"Days off?" asks Anne. "Not really."

"But we also do lots of fun things. We go out to lunch or play in the park. Or we might just stay home and hang around the house. That can be fun, too."

Michael gets home around 6 o'clock. While the boys play outside, Anne and Michael get dinner ready. It's spaghetti and meatballs.

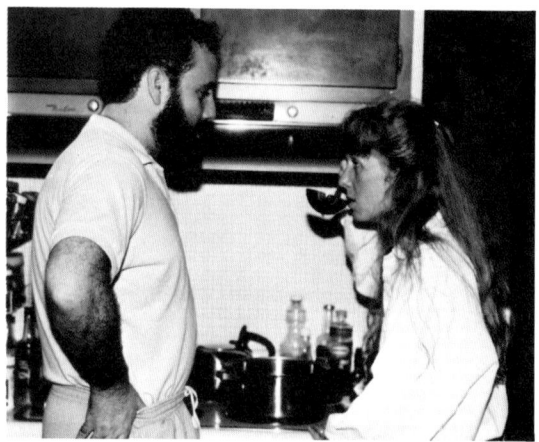

"What should we do this weekend?" Anne asks. Weekends are a special time for the Abrams—a time to slow down and be together.

"If it's warm enough," Michael says, "let's take the boys for a hike in the woods." Several weekends ago, they followed the trail that winds along the Brandywine. Anne and Michael walked slowly so that Alex and Sam could keep up. But even so, Sam got tired quickly, and Michael had to carry him most of the way.

Tonight Anne has more work to do. She has to prepare for the next meeting of the local town council, where she will present the plans for the new housing development. The council must approve the final site plans before work can begin on the project.

The council members are sure to have many questions. Anne's job is to know every detail of the project.

Michael has a big assignment, too. His job is to get the boys ready for bed.

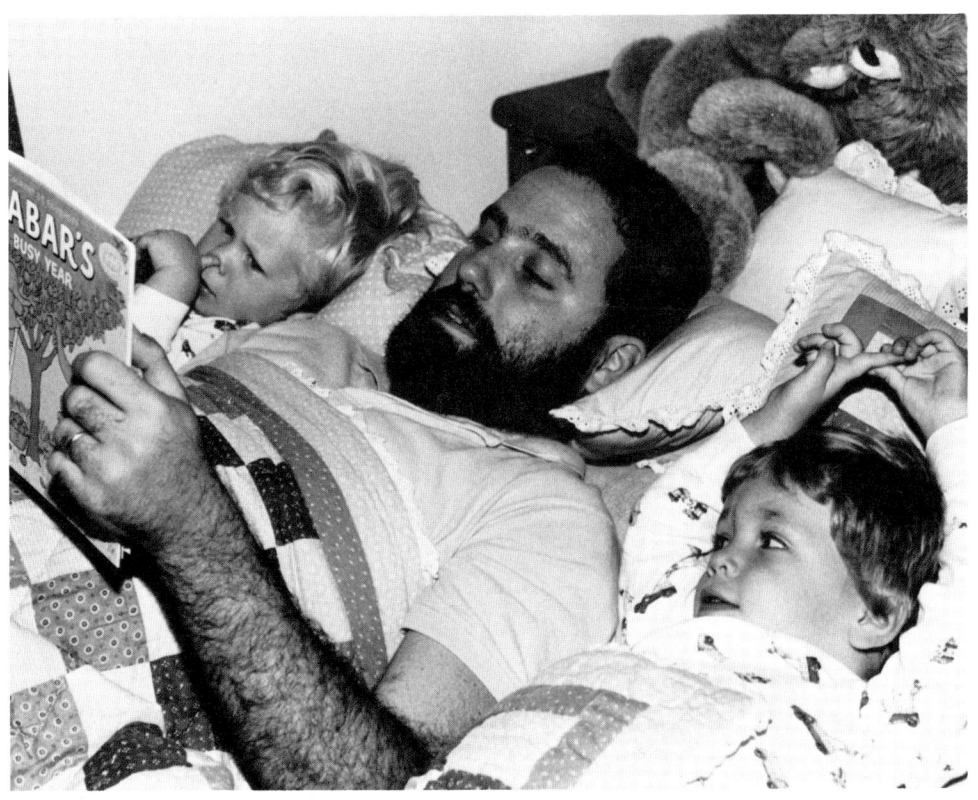

It's 10 o'clock: the end of another long day. Anne puts her drawing instruments away and turns off the lights in her office. She goes back upstairs and looks in on the boys. Michael has been successful: both boys are sound asleep. "They must have been tired," Anne thinks. She tiptoes in and gives each one a kiss on the forehead.

Suddenly, Anne realizes just how tired she is. "It *was* a tough day," she thinks. She walks to the living room and makes herself comfortable on the sofa.

Anne thinks ahead to another busy day tomorrow.

"I'm getting used to busy days," she says to herself. "I guess I'm getting used to being a working mom."

Glossary

blueprint — the final version of a technical drawing

compass — an instrument used to measure and draw circles

drafter — a person who makes technical drawings

drafting table — a large, tilted drawing table used by a drafter

engineer — a person who supervises the building of construction projects

inking pen — a special pen used to make technical drawings

landscaping — the placement of flowers, shrubs, or trees at a building or construction site

mylar — a stiff, plastic film used for technical drawings

protractor — an instrument used to measure and draw angles

scale — a measurement of distance or size used on technical drawings to represent a larger measurement

site — the place where a building or construction project is located

site plan — a technical drawing that shows a view or part of a building or construction project

surveyor — a person who measures the shape of the land and marks off property boundaries

technical drawing — a very accurate and detailed sketch of a building or construction project

transit — a telescope-like instrument used by a surveyor to make a detailed map of a building or construction site